Agricultural Options:

A Primer for Producers

Agricultural Options:

A Primer for Producers

Catania•Polt•Ware

Doane Publishing
11701 Borman Drive • St. Louis, Missouri • 63146

Library of Congress Catalog Card Number: 84-072105
ISBN: 0-932250-21-1

Printed in U.S.A.

Patrick J. Catania

Patrick Catania has over fourteen years experience in the futures industry as an account executive, sales manager and floor trader. He is a frequent speaker on agricultural futures and options at both futures industry and agricultural industry meetings and conventions. In addition, he has developed and taught futures trading classes at several universities. Mr. Catania holds both a Bachelor's and a Master's degree in Accountancy from Western Illinois University.

Julia R. Polt

Julia Polt has been instrumental in the development of numerous seminars on agricultural futures and options for brokerage firms and state and national trade associations. In addition, she has developed literature and audio/visual aids furthering agricultural options education. Ms. Polt holds a Bachelor's degree in Business Administration from William Woods College in Fulton, Missouri, and is currently pursuing her M.B.A. at DePaul University.

John J. Ware

John Ware is currently Marketing Manager for agricultural products at the Chicago Board of Trade. In that capacity he has developed several pieces of literature and published articles on agricultural futures and options in a wide variety of publications. He is also a frequent speaker on agricultural options at many national, state and local conferences on agricultural options. Mr. Ware is currently in his final year in Northwestern University's Executive Masters program. He earned his Bachelor of Arts degree from the University of Minnesota.

Contents

Trading of options on agricultural futures represents the first really new marketing innovation available to farmers in many years.

1 New Promise from An Old Idea

Late in 1984 options on agricultural futures are scheduled to begin trading on several U.S. futures exchanges. What will this mean to you? Primarily, agricultural options will offer the ability to insure a floor price for the sale of your crop without losing the opportunity to participate in rising prices. If you are a livestock feeder or a commercial processor you will be able to insure a ceiling price for the purchase of grain and still benefit from a decline in prices.

Price insurance

The ability to insure either a floor price or a ceiling price is the single most important concept of options. This is a significant departure from traditional futures hedging or forward contracting, where prices are locked in several months in advance. Agricultural options will offer the farmer, the cattleman, the grain dealer and every other sector of agricultural business a tool in developing an effective marketing plan and reducing exposure to price risk.

Yesterday's flaws

While options on agricultural futures are a new concept, the principles of options trading have been around for decades. The first agricultural options were known as *privileges.* They were not formally regulated, were not

1

centrally traded, and were subject to potential manipulation and defaults. This caused Congress in 1936 to place a ban on all domestic agricultural options trading.

Shortcomings remedied

Today options are making a comeback thanks to the efforts of the U.S. exchanges and federal regulatory agencies. Unlike their predecessors, today's options are subject to strict market surveillance practices which have assisted in assuring the integrity of exchange-traded options.

The first step came in 1973 with the founding of a listed stock options market at the Chicago Board Options Exchange. Later came several options on financial futures, including options on U.S. treasury bonds, stock indexes, and currency futures. The success and benefits of these and other options contracts led Congress in 1982 to lift the ban on agricultural options.

The pilot program

Under the Commodity Futures Trading Commission's pilot program, each commodities exchange will be allowed to trade two agricultural options contracts. (A listing of each exchange and their current options contract submissions can be found in the appendix.) These developments have introduced a whole new generation of exchange-traded options supported by the financial integrity of the commodities and securities exchanges.

While an understanding of futures markets is helpful, it is not essential to picking up the options concepts presented here. The beginning sections of this book, which explain what an option is and outline basic options terminology, are absolute essentials if this is your first exposure to options. However, if you already have a basic understanding of what an option is and the language associated with options, you may prefer to begin with the later sections. These sections cover the differences between options and futures, the advantage of using options, and some basic strategies applicable to your operations as producers or feedlot managers.

Wide-ranging uses

Agricultural options hold a great deal of potential for all sectors of agri-business. In addition to those strategies for farmers and producers presented here, there are many more strategies applicable to other commercial agricultural concerns.

One of the major advantages of agricultural options is their potential to assist in securing bank financing. Many banks, previously hesitant to assist the farmer in developing a futures hedging program, will respond favorably to the "price insurance" afforded by options. This fact alone should make the use of options very attractive.

Beginners benefit

While much of the terminology and discussion revolving around options seems complex and difficult, the basic concepts of options trading are relatively simple. It is easy to be scared off by the technical language of options, but in the following pages you will find many simple strategies which you can easily use. You don't need to be an expert in options in order to benefit from these new markets.

Do not, however, conclude that agricultural options are the ultimate solution to all of your marketing problems. Options will not replace current marketing alternatives such as forward contracting, deferred payment, etc. Used effectively, they make up but one part of a comprehensive strategic marketing plan.

Like the traditional land purchase option, an option on an agricultural commodity insures against adverse change in price

2 What is an Option?

A choice

An option, in the simplest sense, is a choice. The purchaser of an option buys the right, but not the obligation, to buy (or to sell) the underlying commodity at a predetermined price, on or before a specified future date. It is the option buyer, though, who has the choice. He can choose to buy (or sell) the commodity, or he can let his option expire, at which point it becomes worthless.

On the other hand, the seller of an option receives an initial payment from the buyer in return for granting him this choice. In this way, options are a unilateral contract binding on the seller only.

A land purchase option

One of the simplest types of options is an option to purchase land. You may be considering purchasing a particular parcel of land to expand your operations. The land in question is a 100-acre tract and the seller is asking $1,000 per acre.

Fee buys time to decide

This price sounds attractive, but you may be unwilling or unable to make a commitment at this point. Instead, you offer to buy an option on the land. It is negotiated between you and the seller that you will pay him a $2,500 fee, and in return you have six months over which to decide if you want to purchase the land. Over this time period the price remains $1,000 per acre.

5

Your decision

1. Exercise your right

2. Sell your option

3. Let option expire

Several things could happen over the course of six months. The price of similar plots of land in the community may go up to $1,200 per acre. This would make the $1,000 per acre purchase price very attractive, and you would choose to exercise the option. But what if you decided not to purchase the land at all? Someone else may be willing to buy the $1,000 per acre option from you (assuming the rights are transferrable) and may be willing to pay more than the initial $2,500 fee.

What if the price of land went down to $800 per acre? Then you'd probably let the option expire worthless rather than pay $200 per acre more than the current market price. In all of these cases, the $2,500 fee (called an option premium) is paid up front to the seller and is nonreturnable.

In the land option example, the option buyer has the right to buy a parcel of land at a predetermined price for a stated period of time. An option such as this which grants the "right to buy" is more commonly referred to as a *call option.*

An option to sell is a *put option.* As a farmer, you may find that this is the type of option which holds the most interest for you. A put option could guarantee months in advance a minimum selling price for your crop. For this reason, options are quite often called *price insurance.* They can be used to insure a floor price or a ceiling price in exchange for the payment of a premium.

Early Ag Options vs. Futures Options

The earliest agricultural options were options to buy or to sell the actual commodity. A soybean call option was just that-the option to buy soybeans. The options discussed in this book, however, are options on agricultural futures contracts. Today a soybean call option would convey to you the right to buy a soybean futures contract.

There are several good reasons for this distinction. The most important reason relates to basis considerations. **Basis** is the difference between current cash prices and the nearby futures contract price. Because the basis relationship remains relatively stable (as compared with actual market prices), it is an important tool in hedging.

Basis

An option on a futures contract will reflect the same basis differential as that of the futures contract. Trading an option on the physical commodity would not facilitate this close relationship because of locational differences, quality differences, etc.

A regulated operation

Secondly, these options will be formally regulated by the exchanges and a governmental unit, the Commodity Futures Trading Commission. They will be subject to the same market surveillance measures currently used to monitor futures markets. The exchanges provide financial integrity through a clearing corporation, which clears all trades and guarantees performance.

The options on physical commodities of the early 1900s did not have these features. Without a formal regulatory body governing trading, there was always the chance that the option seller would renege on this obligation. This will not be the case with exchange-traded options on futures contracts.

Centralized trading

A third reason for exchange-traded options on futures contracts is that they will be traded in a centralized location. This provides two important benefits: accurate pricing (based on supply and demand) and liquidity.

An option on a particular futures contract will be traded only at the exchange which currently trades the underlying futures contract. Futures contracts, through the assembling of buyers and sellers in a competitive central marketplace, facilitate the conversion of supply and demand factors into price. The same will be true for options on futures contracts. Liquidity is also important for a successful option contract because it makes it easier to enter and exit the market. Both of these objectives can be met if the option is traded in a centralized market at a commodity exchange.

*Farming and ranching have their own unique languages, and s[o]
does options trading. Learning the terminology will help you t[o]
grasp the concept[s]*

3 Options Terminology

The first and most important step in gaining a basic understanding of options on agricultural futures is to become familiar with the terminology associated with options. For the first-time reader, several of the terms may seem foreign and confusing. To make matters worse, in many instances several terms will refer to the same concept. For example, to **buy** something could also be termed **going long** or being **bullish.**

These multiple terms for the same concept are widespread in options. However, this section will attempt to take only the most basic terms and clarify their meaning. Other names for the same concept will be pointed out where appropriate.

Buyer and seller

The first step in understanding the mechanics of options is to recognize that every option has a buyer and a seller. If you were the buyer of an option (also frequently called the **holder**) you would have the right, but not the obligation, to exercise that option. On the other hand, the person who sold you that option (often called the **writer** or **grantor**) is required to meet the obligation if you choose to exercise your right. In other words, the buyer of an option has a right, but no obligation. The seller of that option assumes a potential obligation.

Buyer	**Seller**
(holder)	*(writer or grantor)*
Right, but no obligation	Subject to potential obligation

Puts and calls

There are two kinds of options: puts and calls. In the simplest sense, if you bought a call option, you would have the **right to buy.** Buying a put option, on the other hand, would give you the **right to sell.**

In the case of options on agricultural futures, the concept becomes slightly more complex. The right to sell, or a put option, is more accurately described as the right to **go short** in the futures market at a specified price. Call options, or the right to buy, are actually the right to **go long** in the futures market at a specified price.

It is important to understand that the right to buy and the right to sell apply only if you are the purchaser of an option. The person who sells the option is liable to assume a futures market position exactly opposite to that of the buyer's.

It is also very important at this point to recognize that puts and calls are two separate and distinct contracts. Each has its own buyer and seller. They are not opposite sides of the same transaction.

	Buyer	Seller
Calls	Has the right, but not the obligation, to assume a *long* futures position.	Assumes potential obligation to take a *short* futures position.
Puts	Has the right, but not the obligation, to assume a *short* futures position.	Assumes potential obligation to take a *long* futures position.

Underlying contract

Options on agricultural futures are just that: an option on a specific underlying futures contract. For example, your option may be on July corn futures or October live cattle futures. It is important to realize that these options apply to futures contracts and not to the actual physical commodity.

Strike price

The specific price at which the buyer of a call option would assume a long futures position, or the price at which the buyer of a put option would assume a short futures position, is the strike price. (The seller of the option must enter a futures position opposite that of the buyer's.) Another term for strike price is *exercise price,* because it is the price at which you may choose to exercise the option.

Strike prices are predetermined by the exchanges and set at regular price intervals. This allows you the flexibility to choose a strike price that best fits your strategy. Examples of strike prices are July corn futures at $3.50 per bushel or October live cattle futures at $68.00 per cwt.

Exercise

Exercising an option is the action you must take if you are the holder (purchaser) of an option and you want to convert your option to a futures market position. Exercise must take place prior to the option's expiration date.

Expiration

If you are the buyer or holder of an option, the expiration date is the last day on which you may choose to exercise your option. After this date the option is null and void. Options typically expire in the month preceding the futures contract delivery month.

For example, a November option will typically expire in October. It will still be referred to as a November option, though, because its exercise results in a November futures position. This allows time for an orderly liquidation of any futures position which resulted from exercised options.

Liquidation

Besides holding the option until expiration, it is also possible to liquidate your position by selling the same option. For example, if you were holding a November soybean put at $7.50 per bushel, you could liquidate the option by selling a November soybean put at $7.50 per bushel.

You may choose to do this in order to recover some of the initial premium paid. Selling options for this purpose does not create an obligation for the seller because it is done to offset or liquidate a prior transaction, the option purchase.

Premium

The price of an option is called the premium. The premium is paid up front and represents *the maximum amount of potential loss* to which you as an option buyer may be exposed. The option seller, on the other hand, receives this amount when he sells you the option.

As mentioned previously, strike prices and expiration dates are predetermined by the exchanges. Premiums, though, are the one contract term negotiated in the trading pits. How much would you be willing to pay in the spring for the opportunity to sell your soybeans for $8.50 per bushel at harvest? Or how much would you be willing to pay to guarantee a minimum selling price of $6.50 per bushel? This price, or premium, is what will be determined through open outcry in the trading pit.

Using the terminology that has been defined up to this point, let's look at how these terms are combined to describe a particular option . . .

Buy or Sell		Strike	Futures Month	Underlying Contract	Put or Call		Premium
Buy	a	$6.50	July	soybean	put	for	$ .25/bu
Sell	a	$70.00	June	live cattle	call	for	$3.00/cwt

Break-even point

When you purchase an option you must pay a premium. This results in a net debit. In order for you to recover this initial cash outlay, the futures price must move in a favorable direction equal to the premium amount. This point is called your break-even point.

For buying call options, the break-even point occurs when the futures price equals the *initial* strike price *plus* the premium payment. For buying put options, the break-even point is when the futures price equals the initial strike price *minus* the premium payment.

Call Option Breakeven

Futures Price = Strike Price + Premium

Put Option Breakeven

Futures Price = Strike Price - Premium

In-the-money

An option is considered in-the-money when the current futures price is at a point which makes the option profitable for the option holder. A call option would be in-the-money when the futures price is greater than the strike price. This would mean that if you were the holder of a call option you would have the right to buy, or assume a long futures position, at a price below the current market level.

Put options are considered in-the-money when the futures price is lower than the strike price. If you were holding this option you would have the right to sell, or assume a short futures position, at a price above the current market.

CALL OPTION
(CBOT August 1984 soybean futures)

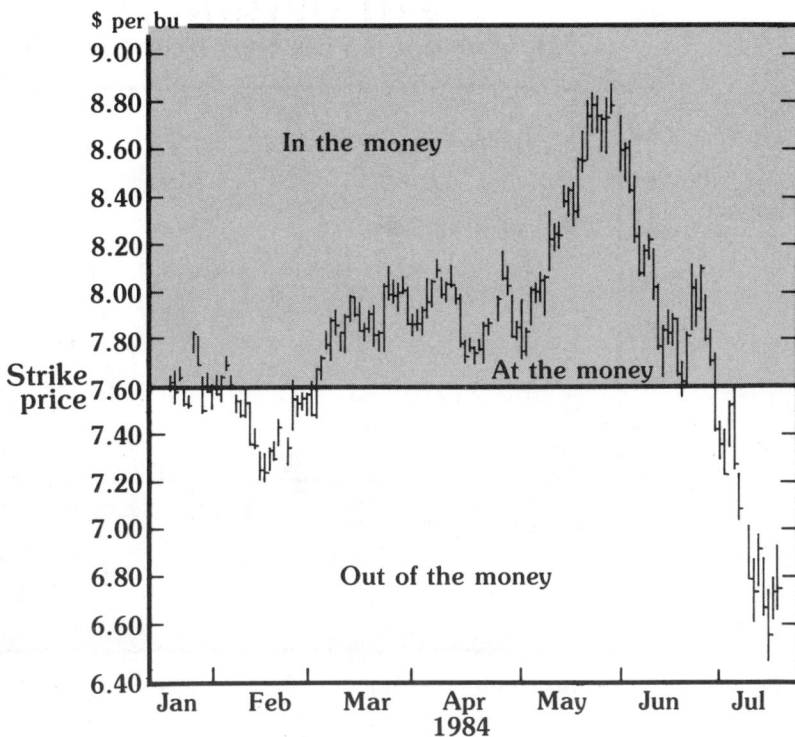

$ per bu
9.00 — 8.80 — 8.60 — 8.40 — 8.20 — 8.00 — 7.80 — Strike price 7.60 — 7.40 — 7.20 — 7.00 — 6.80 — 6.60 — 6.40

In the money

At the money

Out of the money

Jan Feb Mar Apr May Jun Jul
1984

At-the-money

An at-the-money option is one where the futures price is equal, or approximately equal, to the strike price. This option is neither profitable nor unprofitable.

Out-of-the-money

Out-of-the-money options are the opposite of in-the-money options. An option is out-of-the-money when the current futures price is at a point which makes the option holder's position unprofitable. (If the option were exercised, you would receive an unprofitable futures position.)

If you were holding a call option, this would occur when the futures price is lower than the strike price. An out-of-the-money put occurs when the futures price is greater than the strike price.

PUT OPTION
(CME August 1984 live hog futures)

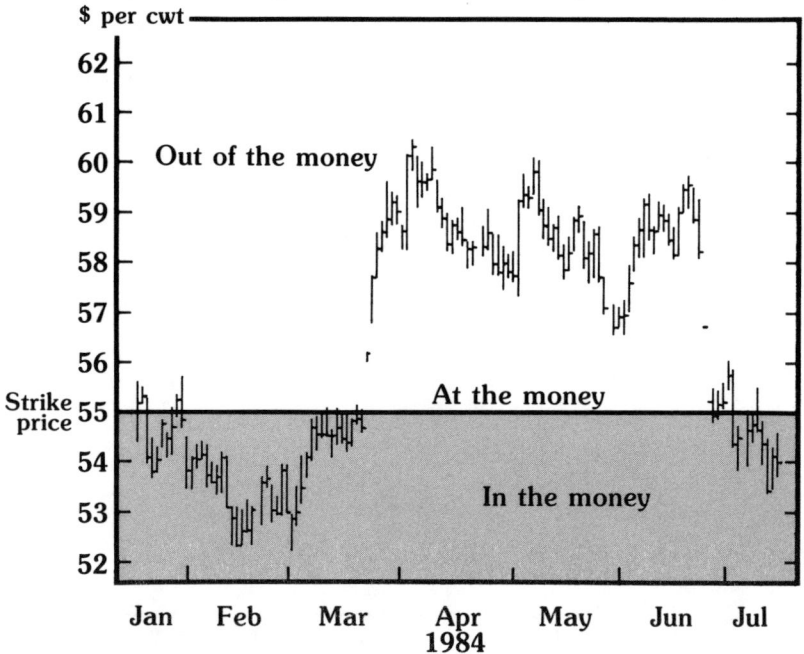

$ per cwt

62
61
60 — Out of the money
59
58
57
56
Strike price 55 — At the money
54
53 — In the money
52

Jan Feb Mar Apr May Jun Jul
1984

These are the basic terms associated with options on agricultural futures contracts. It is difficult to overemphasize the importance of becoming comfortable with this terminology. The next step is to begin to apply some of these terms to actual situations.

*Risk is inherent in the farming business . . . both production
risks and marketing risks. Managing those risks is important
to survival*

4 Risk Management in Farming

Production risk

In the past the producer's primary goal was to maximize production. The entire American agricultural industry focused in on increasing yields and reducing the impact of pestilence and weather-related problems on crop failure. Advances in these areas have given producers many tools with which they can manage their "production risk."

For example, you can choose between very high yield seed corn and seed corn that is, say, wind resistant. There is, however, a trade off. You can't purchase one brand of seed corn that is both the highest yielding and the most wind resistant. (Of course, the seed companies may disagree with this.)

You could purchase the highest yielding seed corn and hope you will get lucky and not experience any high winds during the growing period. Or you could use the extremely wind resistant seed corn and, all else being equal, accept a less than maximum yield. The former choice is nothing more than pure speculation; the latter one might leave you with too small a crop to provide an adequate profit.

By selecting a seed corn with some wind resistance and a good yield potential, you are attempting to gain the best possible trade off. This is a production risk management decision.

21

Price risk

Traditionally, farmers have managed their price risk by using the following tools:

1 Cash sale at harvest

2 Storing at harvest and selling when prices improve

3 Forward contracting at a fixed price

4 Delayed pricing

5 Basis contracting

6 Futures hedging

Each of these pricing alternatives has advantages and disadvantages, and no single pricing method is always the best one to employ under all circumstances.

The use of agricultural options, however, allows you to manage your price risk in ways never before available. You will be able to set a floor (or ceiling) price without giving up the opportunity to take advantage of price increases (or decreases)! Options are a new tool with which you can manage your price risk.

Options and the other previously mentioned tools available to you must all be fully understood in order for you to best manage your price risk exposure. Just as you learn all there is to know about the various brands of seed corn, you must also gain knowledge of all the available pricing alternatives. Only then will you be able to truly manage the entire range of farming risks.

Use of either the options or the futures market provides a means for commodity producers and users to control price risks. But there are important differences in how the two markets work.

5 Options and Futures Compared

A marketing tool

Options will not replace futures hedging or any other marketing alternative. Cash forward contracts, delayed pricing, basis contracts, and cash sales will remain valuable tools. But what options will do is offer you another marketing tool, one that has advantages no other pricing mechanism can match.

Futures hedging

Locking in a price

With a futures hedge, you can substitute an anticipated future cash market purchase or sale by buying or selling futures now. If, for example, in May the futures price for November soybeans (less your local basis) would assure you an attractive profit, you may decide to lock in that price by hedging a portion of your crop at that price. That is, you make a sale (go short) in the futures market. If, in November, prices have fallen, your short futures position profit should have offset (subject to basis variation) the decrease in your local cash price.

Lifting your hedge

You then lift your short futures hedge by buying it back (going long the same amount you went short in May). At the same time, you sell your cash grain in your local market.

Selling your crop

In this example, you have temporarily substituted a cash sale with a futures sale. When it came time to sell your cash grain, you lifted your temporary futures position and sold your grain in the usual manner. Futures let you separate the delivery of your cash grain from the time you price it. In fact, with futures, you can take up to 20 months to price your crop.

How options differ from futures
Options set floor (ceiling)

With futures, you lock in or fix a price (subject to changes in your local basis). Options, on the other hand, allow you to set a floor price (by purchasing puts) or a ceiling price (by purchasing calls) subject, of course, to basis variations.

Margin: a futures drawback

One of the basic differences between futures and options is how they are bought. Futures positions are secured through good faith deposits (*margins*). Every day a futures hedger's gains or losses are added to, or deducted from, his account.

One problem is that if prices increase, you, as a short hedger, must put up large amounts of cash and/or government securities for margin. You may even have to borrow funds to maintain your futures hedge position. A loss in the futures is usually offset by an increase in cash prices. However, you can't realize the higher cash price until you harvest and sell your crop.

Options . . .

1. Margin avoidance

The buyers of options, in contrast to futures, do not have to put up and maintain margins (although option writers, or sellers, must do so). In fact, *the purchaser's maximum risk is limited to the cost of the option (premium).*

2. Staying power

No matter how far prices move against the option holder's position, he does not have to worry about having to put up more money. In effect, then, you have "staying power" during adverse market moves without the worry of margin calls. This fact alone should make it easier for you to sleep at night.

3. Price flexibility

In addition to the staying power you get by using options, you also have the flexibility of setting a ceiling (or floor) price at a level above (or below) the current futures price. Doing this can lower the absolute cost of your premium.

Consider an example

Soybean futures

In May you see that November soybean futures are trading around the $8.50 per bushel level. You also see that a November put option with a strike price of $8.50 carries a premium of, say, 50¢ per bushel. The problem is that futures prices over the last several weeks have been very volatile, having closed both up and down the daily limit. Some people are saying that you'll see "beans-in-the-teens," while others are predicting a severe price drop by harvesttime.

Reason to buy puts

If you forward contract or hedge your grain in the futures market, you protect yourself on the downside but give up all opportunity to participate in any future price appreciation. In this case, options may offer the best risk management alternative. Purchasing puts will give you downside price protection without giving up the opportunity to benefit from future price advances.

How will you proceed?

Set your floor price

You decide that 50¢ is too much to pay for downside protection. However, put options struck at $7.75 per bushel cost only 25¢. This seems to you to be a good balance of price protection and a reasonable cost for that protection. So you purchase the $7.75 put options and set a floor price at that level.

Your net income

If prices at harvest are still at the $8.50 level, your net income after selling your crop and letting your option expire is $8.25 per bushel ($8.50 less the option premium of 25¢). If prices rise to $11.50, your revenue is $11.25. However, if prices fall to $6.50 you net out at $7.50. That is, you sell your cash beans for $6.50 and sell back your put option or exercise it for a short futures position. The $6.50 plus the $1.25, less the 25¢ you originally paid for the options, yields the $7.50.

Opportunity cost

It should be pointed out that should beans at harvest be selling for $7.75 you would be out your 25¢ premium and an "opportunity cost" of 75¢, because you could have hedged or forward contracted back in May at $8.50.

You decide your risk

This opportunity cost can be viewed much like your car insurance deductible. With car insurance, the premium costs are much higher without any deductible than they are with, say, a deductible of $250. The same is true for options. In our example, the premium for $8.50 puts is 50¢, but for the $7.75 puts the cost is only 25¢. You can decide how much downside risk you want to take.

In summary, purchasing options, as opposed to futures, gives you freedom from margin requirements and margin calls, the flexibility to select the level of price protection you want, and the ability to participate in future price increases all for a limited fixed premium that is known and paid up-front.

Prices for both options and the underlying futures contract ar
determined in an open, highly competitive environmen

6 Options Pricing

Determining prices

The price you pay for a put or a call (the premium) is determined in the trading pits of the exchanges in the same way that futures prices are determined. Further, the price of any commodity in a free market system is determined through supply and demand at any given point in time. (Of course, the U.S. and other governments sometimes step in and interfere with the workings of the markets.) The trouble is that there is really no way to calculate a market price.

You will be able to check the previous day's closing premiums in the financial sections of newspapers. The Wall Street Journal, for example, will carry listings like the one below.

CHICAGO BOARD OF TRADE												
Soybeans 5,000 bus. (Cents per bushel)												
Strike price	**Calls**						**Puts**					
	Mar	May	July	Aug	Sep	Nov	Mar	May	July	Aug	Sep	Nov
$5.75	*	*	*	*	*	*	*	*	*	*	*	*
6.00	*	*	*	*	*	*	*	*	*	*	*	*
6.25	*	*	*	*	*	*	*	*	*	*	*	*
6.50	*	*	*	*	*	*	*	*	*	*	*	*
6.75	*	*	*	*	*	*	*	*	*	*	*	*
7.00	*	*	*	*	*	*	*	*	*	*	*	*
7.25	*	*	*	*	*	*	*	*	*	*	*	*
7.50	*	*	*	*	*	*	*	*	*	*	*	*
7.75	*	*	*	*	*	*	*	*	*	*	*	*
8.00	*	*	*	*	*	*	*	*	*	*	*	*

* Each asterisk will be replaced by a premium quoted in cents and eighths of a cent per bushel, for example 25-3/8, meaning 25-3/8¢ per bushel.

Estimating options costs

There are ways, however, to estimate the approximate cost of options. Somewhat complex mathematical formulas must be worked through, but most brokerage houses now have them on computer. In addition, several programs are available for use on your own home computer.

But you will most likely call your broker and ask him or her what the current premiums are for puts or calls at various strike prices. You will then choose the combination of strike price and premium that you feel will give you the desired price protection for the amount you're willing to pay.

What affects costs?

You will benefit if you gain an understanding of the variables that go into the pricing of an option. By doing so and by watching how the options trade, you will develop a "feel" for how premiums change under various market conditions.

Intrinsic value

Basically, options premiums are made up of two parts: *intrinsic value* and *extrinsic,* or *time value.* Intrinsic value is the amount, if any, that the option is in-the-money. In the case of a call, it's the amount the futures price is over the strike price. With a put, it's the amount the futures price is under the strike price.

1. Call option purchase

For example, suppose you purchased a soybean call option with a strike price of $6.50 per bushel and the futures price moved to $7.00 per bushel. What would be the intrinsic value of your call option? Since the futures price is 50¢ above the call option strike price of $6.50, your intrinsic (in-the-money) value would be that 50¢. At the very least, you could sell back or exercise the option for its intrinsic value.

**2. Put option
purchase**

With a put, the situation is reversed. That is, if you owned a $7.00 per bushel put and the underlying futures price was $6.80, your intrinsic value would be 20¢. Here again, your put would be worth at least 20¢ if you were to sell it back or exercise it for a short futures position.

**3. No intrinsic
value for
at-the-money**

Now, if you were to purchase a put or a call struck at-the-money (the futures price and the strike price are the same), the premium you paid would include, logically enough, no intrinsic value. By our definition, intrinsic value only exists by the amount a given put or call is in-the-money. Therefore, any option struck at-the-money has no intrinsic value.

**4. Zero
intrinsic value for
out-of-the-money**

What if you bought a soybean option call at $7.00 and the futures price moved to $6.77? What's your intrinsic value? Here again, the intrinsic value is zero because the option is not in-the-money. In this case, you are actually out-of-the-money.

**5. In-the-
money**

Always remember, intrinsic value is the amount, if any, that a put or a call is in-the-money. An option that is at- or out-of-the-money has zero intrinsic value. Intrinsic value cannot be negative. If an option is out-of-the-money, its intrinsic value is still zero.

Time value

While calculating intrinsic value is relatively easy, understanding an option's time (or extrinsic*) value is somewhat more difficult. There are three major components that make up time value: the time until expiration, interest rates, and volatility.

---

*While extrinsic value is a more definitive term for what is called time value, we will use the latter term because the options industry commonly uses time value.

1. Time till expiration

Everything else being equal, the longer an option has until it expires the more it will cost. This idea is easy enough to understand. If, for example, you were purchasing an auto insurance policy, you would expect one year's coverage to cost more than a six month policy.

The catch is that, for an at-the-money option, the time value component of the premium loses value much more quickly in its last 30 to 45 days than it does up until that time period. For example, the time value of an option with one month until expiration will be less than one-half the time value of an option with two months until expiration.

You should be aware of this factor if you are ever rolling an option hedge into a more deferred month. That is, do not wait until the last few weeks to roll your hedge. Do it before the options time value disappears. The decrease in time value is called **time decay** in options terminology.

One area where you need to account for time decay is in making a decision whether to exercise an option for a futures position or offset it by selling it back. In most cases, you would want to sell an option back if it has more than 30 to 45 days remaining before it expires. If you were to exercise it for a futures position, you would forfeit any remaining time value and only gain the option's intrinsic value.

2. Interest rates

Interest rates also play a role in the time value of options, although not to the extent that time-to-expiration and volatility do. The purchase of an option is an investment, and options must compete with other investments for public and institutional dollars.

When short term interest rates increase, investors' dollars flow into those investments that reflect the higher yield. Options' premiums must, therefore, go down in order to remain a competitive investment. In effect, you could then purchase more options with the same amount of money.

What this all means is that as interest rates rise, all else being equal, options premiums will fall. And, of course, as interest rates fall, options premiums will rise. It is important to note, however, that this interest rate factor has only a minor effect on the value of an option; changes in the other option price variables can easily overshadow or offset interest rate fluctuations.

3. Volatility

The third component of the time value portion of an options premium is volatility. It is relatively easy to determine intrinsic value, time-till-expiration, and the current short term interest rate. Volatility, however, has to be predicted.

Market volatility is a measure of the variability of the underlying futures price. Futures contracts on commodities that fluctuate up and down greatly have higher volatilities than those where prices remain stable.

Soybean futures, for example, are historically much more volatile than corn futures. That would translate into soybean option premiums being relatively more costly than options on corn futures.

a. historical pattern

The problem is that in calculating volatility you are using historical data to predict a current or future volatility. Changes in market supply and demand may alter the historical volatility pattern.

b. risk from fluctuation

Let's say, for example, that cattle futures prices had fluctuated greatly over the last couple of months. Premiums would have to be increased to compensate cattle option writers for the increased risk that the options they sold would move in-the-money at some time in the future. Note, too, that since no one really knows what volatility for a given futures contract will be, each options buyer and seller makes his own best estimate.

The delta factor

The remaining factor that needs to be covered in any discussion of options pricing is the *delta factor*. The delta factor represents the percentage change in an options premium given a change in the underlying futures contract. For example, if a delta of a given soybean call option is 50¢, this implies that the call premium will increase or decrease by a factor of 50% of an advance or decline in the soybean futures price.

Deltas range from near zero for deep out-of-the-money options to 1.0 (100%) for deep in-the-money options. A general rule of thumb is that at-the-money options deltas are about .5. Note that options prices usually are less volatile than futures prices (their deltas are generally under 1.0). Therefore they often will continue trading even after futures have locked up or down the limit.

Deriving fair market premiums

Now that we have discussed the factors that go into determining options prices, it might be beneficial to show you some actual premiums calculated using Professor Black's options pricing model. The table lists "fair market" premiums evaluated for at-the-money, in-the-money, and out-of-the-money puts and calls for various expiration dates. The strike price used in the premium calculation was $7.00, and the short term interest rate was set at 9%. The September 1983 CBOT soybean futures contract was used, and the volatility calculated for that contract was 23.14% over the period from March through May of 1983.

CALCULATED "FAIR MARKET" PREMIUMS FOR SOYBEANS

Days to Expiration	Strike Price	Futures price				
	$7.00	$7.40	$7.20	$7.00	$6.80	$6.60
		(Per bushel)				
30	Call	$ .45	$ .30	$ .18	$ .10	$ .05
	Put	.05	.10	.18	.30	.44
60	Call	.51	.37	.26	.17	.10
	Put	.11	.17	.26	.36	.50
90	Call	.56	.43	.31	.22	.15
	Put	.16	.23	.31	.42	.54
180	Call	.66	.54	.43	.34	.26
	Put	.28	.35	.43	.53	.64
270	Call	.74	.63	.52	.42	.34
	Put	.37	.44	.52	.61	.71
360	Call	.80	.69	.59	.49	.40
	Put	.44	.51	.59	.67	.77

Options can be utilized by an individual or firm that wishes to set a price floor (or ceiling). Farmers should benefit from options even though they may choose not to use them directly

7 Who Will Use Ag Options?

Although we do not yet know precisely who will use ag options, we can certainly predict which individuals and entities will have an interest in this new market. Let's look at these potential users, then look at examples of how ag options might be used.

Commodity producers (farmers)

You, for one, have a vested interest in any mechanism that facilitates the pricing of your production. Whether you use ag options directly or benefit because your local elevator or co-op uses them makes little difference. The fact remains: If you can benefit, you should find a way to use them.

The most obvious direct use that you could make would be to buy put options against your production, as discussed earlier. In the following pages you will be able to work through some examples that show exactly how ag options work and how you could benefit from using them directly.

Food processors

For any food processor, be he soybean processor or meat packer, the acquisition of raw materials (live cattle or soybeans) is a major element of operations. Acquisition costs certainly weigh on bottom line performance, and management closely follows the day-to-day price fluctuations in agricultural markets.

Just as you are able to insure a floor price for your production through the purchase of put options, a processor is also able to insure a ceiling price on his raw materials through the purchase of call options. He is not obligated to perform as the buyer of an option, so if prices fall he can benefit from a lower acquisition cost in the cash market and let his option expire.

Export firms

Exporters often must tender offers to would-be buyers without knowing whether their offer will be accepted. Just as you are at risk with unsold crops in the storage bin or livestock in the feedlot, these exporters are at risk during the time that their offer is extended.

Ag options offer exporters another alternative for managing this type of risk. They can buy call options covering the quantity of exports they have tendered. They know that the options will protect them until their offer is accepted or rejected. At this point they would liquidate their option position, and only then would they go to the cash market if the offer was accepted. Any change in price between the time they extend an offer and the time it is accepted or rejected will be reflected in the value (premium) of the option.

Traders

Traders or speculators will use the options markets just as they use the futures markets. They will buy and sell options in the hope of benefiting from favorable price changes while they hold their position. Their willingness to buy or sell is a willingness to accept the risk you want to transfer.

The types of strategies used by a trader are many and varied. Many involve the use of futures in conjunction with options, and many involve the sale (or writing) of options in addition to the purchase of options. Although most of these strategies are material for a more advanced text on ag options, working through the next section of examples should give you a solid grasp of basic options strategies.

Terminal grain elevators, local elevators, and co-ops

Elevator operators have a great deal of price risk exposure. At harvesttime, when they bid on and buy quantities of grain long after the Chicago futures markets have closed, they are at risk overnight.

Generally, they manage this type of risk with a **prehedge**, or short sale in futures, just prior to the close of the market. Then they even up the following morning, attempting to match actual purchases with sales in the futures market.

The overnight fluctuations in the market price can be a cost to the elevator, one which is passed on to the producer in terms of lower cash bids.

There are ag options strategies that may lower the degree of price risk exposure for the elevator and consequently raise the cash bids to you. Many of these strategies involve the writing (or sale) of options and properly belong in a more advanced book on ag options.

Remember, greater risk exposure increases the cost of doing business. If the elevator operator can minimize risk, he can bid more competitively for your grain.

Livestock feeders

Over and above fixed costs of operation, feed and supplement costs are of primary concern to the livestock feeder. Before you purchase your feeder stock, you can look at the futures markets to determine the cost of corn and soybean meal, as well as the future value of your finished cattle or hogs. If the numbers are "right" you can proceed with the operation, locking in costs and revenues through the futures market.

Call options afford you the opportunity to put a ceiling price on your feed requirements; put options allow you to set a floor price on your finished livestock. And both options still allow you to do better in these areas, should feed prices fall or livestock values rise. The examples which follow will detail just how this works.

By purchasing "call" options, such users of agricultural commodities as cattle feeders can set a ceiling on the cost o.
needed inputs.

Strategies for Using Options

In this section you need a pencil in hand and a pad of paper nearby. We want to cover some possible ways you might use options, as well as some ways the elevator operator and the livestock feeder might do so. Why? Because by seeing how the various users could employ ag options you will get a clearer picture of your own potential uses. Don't hesitate to pencil in current numbers to approximate the outcome of these examples based on more timely data.

Scope of examples

The examples shown here relate only to the purchase of puts or calls. The sale of puts or calls involves the granting of an option and is not subject to the limited risk of buying an option as discussed earlier.

These examples will, however, give a clear picture of the potential of ag options. They will point out that ag options are not intended to replace a well defined risk management and marketing program. Rather, they are designed to enhance such a program.

Also, the examples will use the nearby futures contract to approximate the current value of your production, ignoring basis. This will require you to allow for the basis in determining the actual outcome.

Alternatives

Bear in mind that when you buy an option on agricultural futures (put or call) you have three alternatives as to the disposition of that option:

1. Sell the option back into the marketplace:

Assume that you buy a November $7.50 put option on soybean futures and pay 25¢ per bushel for that option.

If November soybean futures go down, the value of the option (25¢ per bushel premium you paid) could go up. Hence, if an option you bought for 25¢ per bushel is now worth 30¢ per bushel, you could sell the option for a 5¢ per bushel profit, or $250 (30¢ - 25¢ = 5¢ X 5,000 bushels per contract = $250).

2. Exercise the option and take a futures position:

Given the same circumstances as in the previous alternative, you could exercise your $7.50 November put option, which would give you a short position in November bean futures at a price of $7.50, regardless of where November bean futures are currently priced.

A word of caution: If you exercise an option and take a futures position, you no longer have the benefit of limited risk. Once you take the futures position you are exposed to and can suffer the consequences of adverse price movement on the futures contract.

Of course, if you have the actual commodity, this places you in the position of a traditional futures hedge. Once you have the futures position, any profit (or loss) will be realized upon offsetting that position.

3. Abandon the option and let it expire worthless:

Assume that, as in the previous examples, you buy a $7.50 put option on November beans for 25¢ per bushel. By the time the option expires (approximately the middle of the month prior to the underlying futures contract expiration), November beans are trading at $9.00 per bushel.

You certainly wouldn't exercise your option for a short position (put) at $7.50 with the market currently at $9.00! So you abandon the option with a loss of 25¢ per bushel (the premium you paid), and you sell your beans in the cash market.

The 25¢ per bushel loss on your option would be offset many times over by the much higher current cash price available for your beans. Most importantly, your total risk exposure was limited to the 25¢ per bushel premium payment. If the market went down, you were protected; if it went up you were not locked into a forward sale or futures contract.

Put option strategies . . .

Since the put option appears to be an extremely useful tool for the producer in managing risk, let's look at several examples of puts involving soybeans, and then live cattle. The option on soybean futures is scheduled to be the first ag option offered by the Chicago Board of Trade, while the option on live cattle futures will be the first available at the Chicago Mercantile Exchange.

Remember that all of the premium costs and product values used are hypothetical, and it will be necessary for you to insert actual values into these examples when options trading begins.

Grain and soybean producer: *floor price insurance*

Scenario: It is nearing harvesttime, and you are interested in insuring a floor price for your soybean production. You decide to buy put options covering 10,000 bushels of anticipated harvest. You generally sell after the first of the year, so you are interested in options on March soybean futures. March futures are trading at $8.00 per bushel, and you want to insure a price of at least $7.50 per bushel (based on the March futures price).

Date: September 1

Action: You buy two March $7.50 soybean puts, paying a premium of 25¢ per bushel.

Results: Each put option covers a 5,000-bushel soybean futures contract, therefore you have purchased

the right (but not the obligation) to
short positions totaling 10,000
bushels at a price of $7.50. Your
cost is 25¢ per bushel, or $2,500
(25¢ X 10,000 bushels = $2,500).

What if the price of soybeans goes down by the time you decide to market your crop in January?

Date: January 10

Change: March soybean futures are now
trading at $6.50 per bushel (down
from the $8.00 level on September
1). Your local cash market has
fallen a corresponding amount. You
can either exercise your $7.50 put
option or sell it back into the
market.

Action A: You decide to exercise your
option.

**Exercising
your
option**

Results: If you exercise your option, you will
receive two short March futures
contracts at $7.50 while the current
futures market is at $6.50. You off-
set your short positions and you
make $1.00 per bushel on the
futures contracts. If you deduct the
original 25¢ per bushel premium
you paid for the option, your net
profit is 75¢ per bushel:

50

Exercise option, receive short
futures position at: $7.50/bu

Cover short futures position at
current market level of: −6.50/bu

Gross proceeds paid
from transaction: 1.00/bu

Less premium paid for option: −.25/bu

NET PROCEEDS: $ .75/bu
or $7,500 on 10,000 bushels.

Action B: You decide to sell your option back
into the market.

**Selling
your
option**

Results:
Current value of put option
($7.50 strike price, market
at $6.50) $1.00/bu*

Purchase price of option: − .25/bu

NET PROCEEDS: $ .75/bu
or $7,500 on 10,000 bushels.

Benefit   In either case, the net proceeds from the option transaction would go a long way to offset
the decreased value of your crop in the local
cash market.

---

*The $7.50 put has "intrinsic" value of $1.00 and may have additional "time" value,
depending on its remaining life (date of expiration).

What if the price of soybeans goes up by the time you decide to market your crop in January?

Date: January 10

Change: March soybean futures are now trading at $9.50 per bushel (up from the $8.00 level on September 1). Your local cash market has risen a corresponding amount, and you want to take advantage of the current price.

Action: You simply sell your cash beans in the local market, and abandon your option.

Abandoning your option

Results: You can sell your crop at the current higher price with no lost opportunity as is incurred with a forward contract or a short futures position. Your only cost is the amount of premium you paid for your option, in this case 25¢ per bushel, or $2,500 on 10,000 bushels. This premium payment is more than offset by the $1.50 per bushel rise in the price of beans. On 10,000 bushels, here are the results:

10,000 bu X $1.50 (increase in market value of soybeans) = $15,000

Less cost of premium
(25¢ X 10,000 bushels):   −2,500

NET BENEFITS $12,500

Benefit

With a forward contract or a futures contract you would have been locked into the lower price, forgoing the opportunity to profit from such an increase. In this case, you would have incurred an opportunity cost of $12,500!

Livestock feeder: *floor price insurance*

Scenario: It's early spring and you move 120 head of feeder cattle onto the feedlot. Expected marketing date is next November. Currently, December live cattle futures are trading at $67.00 per cwt. Based on your calculations, you need to sell your cattle at $63.00 per cwt or better. You decide to purchase put options on live cattle futures (at the Chicago Mercantile Exchange) to insure your pricing.

Date: March 1

Action: You buy three December $65 puts on live cattle futures. Each put option grants the right to a short position in December live cattle futures at $65.00 per cwt. Each futures contract covers 40,000 lbs of live cattle (approximately 40 head), so you buy three put options to cover your 120 head. Let's say you pay a premium of $2.00 per cwt for these options. What is the result?

Results: Remember, it's March 1 and December live cattle futures are trading at $67.00 per cwt. You pay $2.00 per cwt for the put options, which works out to $2,400 ($2.00 per cwt X 40,000 lbs X 3 options contracts). You have accomplished your goal of insuring a price of $63.00 per cwt for your cattle because no matter how low the market goes, your options entitle you to sell at the strike price of $65.00 per cwt.

What happens if the price of live cattle futures drops by $6.00 per cwt?

Date: November 1

Change: You're ready to sell your cash cattle. Since the market has dropped by $6.00 per cwt, December futures are now trading at $61.00 per cwt. You have two alternatives: exercise your $65.00 put options, or sell them back into the marketplace.

Action A: You decide to exercise your option.

Exercising your option

Results: If you choose to exercise, you'll receive three short December live cattle futures contracts at $65.00 per cwt. Since you are selling your cash cattle, you will simultaneously want to cover your short futures positions.

Short futures positions:	$65.00/cwt
Cover at current futures price:	-61.00/cwt
Proceeds from exercise and offset:	4.00/cwt
Premium paid for options:	-2.00/cwt
NET PROCEEDS:	$2.00/cwt
Cash price for cattle:	+61.00/cwt
NET EFFECTIVE SALE PRICE:	$63.00/cwt

Action B: You decide to sell your options back into the market.

Selling your option

Results:

Selling price of $65.00 put: $4.00/cwt*

Original premium paid: −2.00/cwt

NET PROCEEDS: $2.00/cwt

Cash cattle price: +61.00/cwt

NET EFFECTIVE SALE PRICE: $63.00/cwt

Benefit

In either case, the objective was met. The net effective price received for your live cattle met your initial objectives.

What happens if the price of live cattle futures goes up by $6.00 per cwt?

Date: November 1

Change: You're ready to sell your cash cattle and the market price is currently $73.00 per cwt. (March 1 price of $67.00 + $6.00 per cwt market increase).

Action: You sell your cattle in the cash market and let your option expire worthless. You would not exercise the right to a short position at $65.00 per cwt ($65.00 put) when the market is trading at $73.00 per cwt! The $2.00 per cwt that you paid in premium for your put option was a "cost of insurance."

Abandoning your option

---

*The $65.00 put options will have intrinsic value of $4.00 because the underlying futures contracts are currently trading at $61.00 per cwt. There could be additional "time" value, depending on expiration date.

Cash cattle price:	$73.00/cwt
Less cost of put options:	-2.00/cwt
NET EFFECTIVE SALE PRICE:	$71.00/cwt

Benefit

You far surpassed your objective of $63.00 per cwt, and you had price insurance over the course of time the cattle were on feed. Remember, with a short futures contract or a guaranteed price from a meat packer, you would have been locked into the lower market price and would have incurred an opportunity cost.

Elevator operator: *prehedge anticipated purchases*

Scenario: It's harvesttime and you are faced with the prospect of buying thousands of bushels of grain. You need to protect yourself on a day-to-day basis, especially for that grain you buy after the Chicago Board of Trade closes each day. Put options offer another alternative as a prehedge for anticipated purchases by a country elevator.

Let's look at the purchase of put options in anticipation of buying 50,000 bushels of corn overnight. December futures are trading at $3.50 per bushel.

Date: October 15

Action: You buy ten December $3.50 put options (each covers 5,000 bushels) at 10¢ per bushel.

Results: You have protection for up to 50,000 bushels of corn bought overnight, and you are able to bid on corn based on the "insurance" you bought with put options.

What happens if corn goes down overnight?

	Date:	October 16

Selling your option

Change: December corn futures open 5¢ lower at $3.45, therefore the corn you bought overnight is moved in the cash market at 5¢ lower. However, you can turn around and sell your put options to recover all or part of the 5¢ per bushel decrease, because a put option increases in value as the underlying commodity price goes down.

Action: You decide to sell your put options.

Benefit

Results: You sell ten December $3.50 put options at 15¢ per bushel, gaining 5¢ per bushel to offset the 5¢ per bushel lost on overnight price action.

What happens if corn goes up overnight?

Date: October 16

Selling your option

Change: December corn futures open 5¢ higher at $3.55, and you move the corn bought overnight at 5¢ higher.

Action: You must also sell your put options, possibly at a 5¢ lower level. The 5¢ per bushel gain in cash corn offsets the 5¢ per bushel loss in the options position.

Benefit

Results: You have achieved protection on your cash market exposure, the cost of which was offset by the increase in the cash market price.

A key point to remember in this example is that if the price of corn were to go sharply higher overnight (USDA crop report, foreign sales, etc.), the put option could be sold or abandoned and you would have no obligation to sell the grain at any set price. If you sell futures short as a prehedge for the elevator, you must offset the futures position or deliver the corn at some point in time.

Call option strategies . . .

We've discussed some potential uses for put options by the producer, the livestock feeder, and the elevator operator. Now let's look at how call options can benefit certain users.

Livestock feeder: *setting cost ceiling*

Scenario: You have a hog operation which requires the purchase of 10,000 bushels of corn on a quarterly basis. Currently prices for corn futures over the next year are attractive, based on the fact that your production of hogs is hedged and your cost analysis shows profit margins can be maintained if feed costs are controlled. You want to lock in a "ceiling price" for your feed requirements without giving up the opportunity to profit from a drop in feed costs.

You decide to buy call options to cover your anticipated needs for the next quarter.

Date: April 1

Action: You buy two July $3.50 call options (July futures currently trading at $3.40) for 10¢ per bushel.

Results: You have insured your acquisition cost of corn will not exceed $3.60 per bushel (the strike price of the option, $3.50, plus the premium paid, 10¢ per bushel, equals the maximum cost of your corn).

What happens if corn prices rise to $3.75 per bushel?

Date: June 15

Change: You're ready to purchase your quarterly feed requirements of 10,000 bushels. You buy your corn in the cash market at $3.75, and either exercise or sell your call options.

Exercising your option

Action A: You decide to exercise your call option.

Results: If you exercise your options, you receive 10,000 bushels of July futures long at $3.50 per bushel. Since the market is trading at $3.75, you would sell your futures contracts and achieve the following:

Sell 10,000 July corn futures:	$3.75/bu
Less purchase price (from exercise of call option):	-3.50/bu
Proceeds:	.25/bu
Less cost of option:	-.10/bu
NET PROCEEDS:	$ .15/bu

Benefit

This 15¢ per bushel reduces your cost of cash corn from the $3.75 per bushel paid to a net cost of $3.60, thus achieving your objective.

Action B: You decide to sell your call
options.

**Selling
your
option**

Results:

Sale of $3.50 call (current market
at $3.75): $ .25*

Cost of $3.50 call: – .10

NET PROCEEDS: $ .15

Benefit

The same result is achieved, reducing net cost
of feed corn from $3.75 per bushel to $3.60
per bushel.

What happens if the price of corn falls?

**Abandoning
your
option**

If corn prices fall by the time you need to
make your purchases, you simply take advantage of the lower cash prices. You are under
no obligation (as the buyer of the call option)
to take corn at the $3.50 strike price. If corn
is currently $3.40 per bushel or $3.20 per
bushel, you let your option expire and pay
the lower cash price.

It should be understood that the cost of your
premium for the call option is a cost of doing
business. The 10¢ per bushel paid in this example must be considered as part of the acquisition cost of your corn. Therefore, if cash
corn costs you $3.20 per bushel and your
$3.50 call option is now worthless, the 10¢
per bushel premium you paid effectively raises
your cost to $3.30 per bushel ($3.20 per
bushel cash price + 10¢ per bushel premium
paid).

*Once again we ignore any "time" value and base the example on "intrinsic" value.

Grain and soybean processors:

price insurance

There are a number of strategies, some very involved, that are applicable to processors and crushers. The basic element of "price insurance" is also available to these market participants.

Grain and soybeans are raw materials to these processors, just as live cattle and live hogs are the raw materials for the meat packer. Very basically, call options afford these entities the ability to lock in or insure a ceiling price for the acquisition of their raw materials.

Also, these entities could buy put options against their production, insuring a floor price. This would afford a total risk management package and enhance the overall return on investment to these firms.

*The options market provides another tool for farmers. But i
is not a panacea. A good marketing program will make use o
other methods, too*

9 Summary

At this point, you should have a basic understanding of options on agricultural futures. We can't stress enough the importance of the basic terminology and concepts to the understanding of more involved strategies.

In summary, a brief overview of the advantages and disadvantages of agricultural options is in order.

Advantages

1. The option buyer faces no margin calls while he holds his put or call.

2. Options allow price protection on growing crops without commitment to a forward contract or short futures position.

3. The maximum financial risk to the buyer of an option is the premium paid.

4. Options provide floor prices (put options) or ceiling prices (call options) in times of uncertainty.

5. The option buyer retains the opportunity to benefit from favorable price changes while insuring against unfavorable price changes.

63

Disadvantages

1. The premium paid by an option buyer is a cost of doing business, and as such is a draw on net returns.

2. Option buyers must pay the full premium at the time of purchase.

3. The buyer of an option holds an "eroding" asset, since time value is a major component of premium cost.

4. Option sellers or writers must employ complex strategies to avoid the unlimited risk of granting the option.

Appendix A
Proposed Contract Terms

On the next few pages are some of the terms of the various commodity options contracts as they stand at the time this book is published. But all terms are subject to change by the exchange where the commodities are traded.

Therefore, refer to the following tables for examples of how terms are set up, but not for reliable information about what may now be in effect. Consult Appendix B for contacts at the exchanges. These contacts or a commodity broker will be your source of current information.

Chicago Board of Trade
Options on Soybean Futures

Trading unit

One (1) Chicago Board of Trade 5,000 bushel soybean futures contract.

Quotation

In dollars and cents per bushel; in minimum increments of 1/8¢ per bushel, half that of the underlying futures contract.

Trading hours

9:30 a.m. to 1:15 p.m. Central Time.

Contract months

Options may be exercised for soybean futures which expire in the months of January, March, May, July, August, September, and November.

Last trading day

Options on soybean futures cease trading on the first Friday which is at least 10 business days prior to the first notice day in the underlying soybean futures contract.

Expiration

Options expire at 10 a.m. on the Saturday following the last trading day.

Strike prices

Strike prices are set at 25¢ intervals.

Chicago Board of Trade
Options on Corn Futures

Trading unit

One (1) Chicago Board of Trade 5,000 bushel futures contract.

Quotation

In dollars and cents per bushel; in minimum increments of 1/8¢ per bushel, half the underlying futures contract.

Trading hours

9:30 a.m. to 1:30 p.m., Central Time.

Contract months

Options may be exercised for corn futures that expire in the months of March, May, July, September, and December.

Last trading day

Options on corn futures cease trading on the first Friday which is at least 10 days prior to the first notice day in the underlying corn futures contract.

Expiration

Options expire at 10 a.m. on the Saturday following the last trading day.

Strike price

Strike prices are established at 10¢ intervals.

Chicago Mercantile Exchange
Options on Live Cattle Futures

Trading unit

One (1) Chicago Mercantile Exchange 40,000 pound live cattle futures contract.

Quotation

In dollars and cents per pound; in minimum increments of .025¢ per pound, same as the underlying futures contract.

Trading hours

9:05 a.m. to 1:00 p.m. Central Time.

Contract months

Options may be exercised for live cattle futures that expire in the months of February, April, June, August, October, and December.

Last trading day

Options on live cattle futures cease trading on the fourth business day prior to the first business day of the underlying future's delivery month.

Expiration

Options expire on the fourth business day prior to the first business day of the underlying future's delivery month.

Strike prices

Strike prices are established at 2¢ intervals.

Chicago Mercantile Exchange
Options on Live Hog Futures

Trading unit One (1) Chicago Mercantile Exchange 30,000 pound live hog futures contract.

Quotation In dollars and cents per pound; in minimum increments of .025¢ per pound, half the underlying futures contract.

Trading hours 9:20 a.m. to 1:00 p.m. Central Time.

Contract months Options may be exercised for live hog futures that expire in the months of February, April, June, July, August, October, and December.

Last trading day Options on live hog futures cease trading on the fourth business day prior to the first business day of the underlying future's delivery month.

Expiration Options expire on the fourth business day prior to the first business day of the underlying future's delivery month.

Strike prices Strike prices are established at 2¢ intervals.

Mid-America Commodity Exchange
Options on Soybean Futures

Trading unit

One (1) Mid-America Commodity Exchange 1,000 bushel soybean futures contract.

Quotation

In dollars and cents per bushel; in minimum increments of 1/8¢ per bushel, same as the underlying futures contract.

Trading hours

9:30 a.m. to 1:30 p.m. Central Time.

Contract months

Options may be exercised for soybean futures that expire in the months of January, March, May, July, August, September, and November.

Last trading day

Options on soybean futures cease trading on the first Friday which is at least 10 business days prior to the first notice day in the underlying futures contract.

Expiration

Options expire at 10 a.m. on the Saturday following the last trading day.

Strike prices

Strike prices are established at 25¢ intervals.

Mid-America Commodity Exchange
Options on Wheat Futures

Trading unit

Five (5) Mid-America Commodity Exchange 1,000 bushel wheat futures contracts.

Quotation

In dollars and cents per bushel; in minimum increments of 1/8¢ per bushel, same as the underlying futures contract.

Trading hours

9:30 a.m. to 1:30 p.m. Central Time.

Contract months

Options may be exercised for wheat futures that expire in the months of March, May, July, September, and December.

Last trading day

Options on wheat futures cease trading on the first Friday which is at least 10 days prior to the first notice day in the underlying wheat futures contract.

Expiration

Options expire at 10 a.m. on the Saturday following the last trading day.

Strike prices

Strike prices are set at intervals of 10¢ per bushel.

Kansas City Board of Trade
Options on Wheat Futures

Trading unit One (1) Kansas City Board of Trade 5,000 bushel wheat futures contract.

Quotation In dollars and cents per bushel; in minimum increments of 1/2¢ per bushel, half the underlying futures contract.

Trading hours 9:30 a.m. to 1:15 p.m. Central Time.

Contract months Options may be exercised for wheat futures that expire in the months of March, May, July, September, and December.

Last trading day Options on wheat futures cease trading on the first Friday which is at least 10 business days prior to the first notice day in the underlying wheat futures contract.

Expiration Options expire at 10 a.m. on the Saturday following the last trading day.

Strike prices Strike prices are set at intervals of 10¢ per bushel.

Minneapolis Grain Exchange
Options on Wheat Futures

Trading unit	One (1) Minneapolis Grain Exchange 5,000 bushel wheat futures contract.
Quotation	In dollars and cents per bushel; in minimum increments of 1/8¢ per bushel, same as the underlying futures contract.
Trading hours	9:30 a.m. to 1:15 p.m. Central Time.
Contract months	Options may be exercised for wheat futures that expire in the months of March, May, July, September, and December.
Last trading day	Options on wheat futures cease trading on the first Friday which is at least 10 business days prior to the first notice day in the underlying wheat futures contract.
Expiration	Options expire at 10 a.m. on the Saturday following the last trading day.
Strike prices	Strike prices are set at intervals of 10¢ per bushel.

New York Cotton Exchange
Options on Cotton Futures

Trading unit	One (1) New York Cotton Exchange 50,000 pound cotton futures contract.
Quotation	In dollars and cents per pound; in minimum increments of 1/100¢ per pound, half the underlying futures contract.
Trading hours	10:30 a.m. to 3:00 p.m. Eastern Time.
Contract months	Options may be exercised for cotton futures that expire in the months of March, July, October, and December.
Last trading day	Options on cotton futures cease trading on the first Friday of the month prior to the delivery month for the underlying cotton contract.
Expiration	Options expire at noon on the Saturday following the last trading day.
Strike prices	Strike prices are set at 2¢ intervals for two deferred months, and at 2¢ intervals for two near months when the underlying future is more than $74.50; when the underlying future is less than $74.50, strike prices are at 1¢ intervals for the two near months.

Appendix B

Commodity Exchanges
Literature Contacts & Contract Submissions

Contact for Literature	Current Option Contract Submissions
Marketing Literature Chicago Board of Trade 141 W. Jackson Blvd. Chicago, IL 60604 (312)435-3558	Soybeans Corn
Office Services Dept. Chicago Mercantile Exchange 30 S. Wacker Drive Chicago, IL 60606 (312)930-8210	Live cattle Live hogs
Market Development Kansas City Board of Trade 4800 Main Street Kansas City, MO 64112 (816)753-7500	Hard red winter wheat
Marketing Development MidAmerica Commodity Exchange 444 W. Jackson Blvd. Chicago, IL 60606 (312)341-3052	Soybeans Soft winter wheat
Marketing/Public Relations Minneapolis Grain Exchange 150 Grain Exchange Bldg. Minneapolis, MN 55415 (612)338-6212	Hard spring wheat
Marketing Department New York Cotton Exchange Commodity Exchange Center 4 World Trade Center New York, NY 10048 (212)938-2702	Cotton

Glossary of Terms

at-the-money   an option whose strike price is equal-or approximately equal-to the current market price of the underlying futures contract.

basis   the difference between current cash prices and the nearby futures contract price.

bearish   a market view which looks toward lower prices.

break-even point   a futures price point at which a given strategy is neither profitable nor unprofitable.

bullish   a market view which looks toward higher prices.

buyer   the purchaser of an option, either a call option or a put option. The buyer may also be referred to as the option holder. Option buyers receive the right, but not the obligation, to enter a futures market position.

call option	an option which gives the option buyer the right to purchase (go long) the underlying futures contract at the strike price on or before the expiration date.
CBOT	the Chicago Board of Trade.
CFTC	the Commodity Futures Trading Commission.
CME	the Chicago Mercantile Exchange.
closing transaction	see *Liquidation.*
commission	fees paid to the broker for execution of an order.
exercise	the action taken by the holder of a call if he wishes to purchase the underlying futures contract or by the holder of a put if he wishes to sell the underlying futures contract.
exercise price	same as strike price.
expiration date	the last date on which the option may be exercised. Although options expire on a specified date during the month prior to the named month, an option on a November futures contract is referred to as a November option, since its exercise would lead to the creation of a November futures position.
extrinsic value	same as time value.

futures contract a contract traded on a futures exchange for the delivery of a specified commodity at a future time. The contract specifies the item to be delivered and the terms and conditions of delivery.

futures price the price of a particular futures contract is determined by open competition between buyers and sellers on the trading floor of a commodity exchange.

grantor see *seller.*

hedge the buying or selling of offsetting positions in order to provide protection against an adverse change in price. A hedge may involve having positions in the cash market or the futures market or holding options.

holder see *buyer.*

in-the-money a *call* is in-the-money if its strike price is below the current price of the underlying futures contract (i.e., if the option has intrinsic value). A *put* is in-the-money if its strike price is above the current price of the underlying futures contract (i.e., if the option has intrinsic value).

intrinsic value the dollar amount which would be realized if the option were to be exercised immediately. See **in-the-money.**

KCBT the Kansas City Board of Trade.

liquidation	a purchase or sale which offsets an existing position. This may be accomplished by selling an option which was previously purchased or by buying back an option which was previously sold.
long	the position which is established by the purchase of a futures contract or an option (either a call or a put) if there is no offsetting position.
margin	in commodities, an amount of money deposited to insure performance of an obligation at a future date. Buyers of options do not post margin since their risk is limited to the option premium, which is paid in cash when the option is purchased.
margin calls	additional funds which a person with a futures position or the writer of an option may be called upon to deposit if there is an adverse price change or if margin requirements are increased. Buyers of options are not subject to margin calls.
MidAm	the Mid-America Commodity Exchange.
MGE	the Minneapolis Grain Exchange.
naked writing	writing a call or a put on a futures contract in which the writer has no opposite cash or futures market position. This is also known as *uncovered writing*.
NYCE	the New York Cotton Exchange.

open interest	total number of futures or options (puts and calls) contracts outstanding on a given commodity.
opening transaction	a purchase or sale which establishes a new position.
out-of-the-money	a put or call option which currently has no intrinsic value. That is, a call whose strike price is above the current futures price or a put whose strike price is below the current futures price.
premium	the price of an option, not including related brokerage commission fees. The premium is the maximum amount of potential loss to which the option buyer may be subject.
privileges	an early form of agricultural option, no longer traded.
put option	an option which gives the option buyer the right to sell (go short) the underlying futures contract at the strike price on or before the expiration date.
seller	also known as the *option writer* or *grantor.* The seller of an option is subject to a potential obligation if the buyer chooses to exercise the option.
short	the position created by the sale of a futures contract or option (either a call or a put) if there is no offsetting position.
strike price	the price at which the holder of a call (or put) may choose to exercise his right to purchase (or to sell) the underlying futures contract.

time value   any amount by which an option premium ex-
ceeds the option's intrinsic value. If an option
has no intrinsic value, its premium is entirely
time value.

underlying futures   the specific futures contract that may be
contract   bought or sold by the exercise of an option.

writer   see **seller.**